ANTIQUES AND THEIR VALUES

MILITARIA

Compiled and Edited by

TONY CURTIS

All photographs and text of the arms were
provided by Wallis and Wallis, 210 High Street,
Lewes, Sussex.

1SBN 0 902921 49 5

Copyright © Lyle Publications l976

Published by Lyle Publications Glenmayne Galashiels Selkirkshire Scotland

CONTENTS

Printed by APOLLO PRESS LTD.,
Unit 5, Dominion Way, Worthing,
Sussex.

INTRODUCTION

Congratulations! You now have in your hands an extremely valuable book. It is one of a series specially devised to aid the busy professional dealer in his everyday trading. It will also prove to be of great value to all collectors and those with goods to sell, for it is crammed with illustrations, brief descriptions and valuations of hundreds of antiques.

Every effort has been made to ensure that each specialised volume contains the widest possible variety of goods in its particular category though the greatest emphasis is placed on the middle bracket of trade goods rather than on those once - in - a - lifetime museum pieces whose values are of academic rather than practical interest to the vast majority of dealers and collectors.

This policy has been followed as a direct consequence of requests from dealers who sensibly realise that, no matter how comprehensive their knowledge, there is always a need for reliable, up-to-date reference works for identification and valuation purposes.

When using your Antiques and their Values to assess the worth of goods, please bear in mind that it would be impossible to place upon any item a precise value which would hold good under all circumstances. No antique has an exactly calculable value; its price is always the result of a compromise reached between buyer and seller, and questions of condition, local demand and the business acumen of the parties involved in a sale are all factors which affect the assessment of an object's 'worth' in terms of hard cash.

In the final analysis, however, such factors cancel out when large numbers of sales are taken into account by an experienced valuer, and it is possible to arrive at a surprisingly accurate assessment of current values of antiques; an assessment which may be taken confidently to be a fair indication of the worth of an object and which provides a reliable basis for negotiation.

Throughout this book, objects are grouped under category headings and, to expedite reference, they progress in price order within their own categories. Where the description states 'one of a pair' the value given is that for the pair sold as such.

Austrian mid-19th Century sword bayonet,
blade 24ins.   £22

A scarce Australian model 1943, S.M.L.E.,
machete bayonet.   £35

A Nazi police parade bayonet by Weyersberg
Kirskbaum, plated blade 13ins.   £46

A brass hilted Prussian 1871 Model bayonet,
slightly curved blade 18½ins.   £64

A rare, Swiss Model 1864 sabre bayonet,
fullered Yatagan blade 20¾ins.   £105

A fine, desirable and important, late 17th Century
hallmarked silver mounted plug bayonet, slightly
curved clipped-back blade 13¾ins.   £550

A late 17th Century Cavalry broadsword, straight single edged blade 34ins. £65

A French cuirassier's broadsword, straight single edged blade 38in. £75

Spanish mid-18th Century cup hilted broadsword, double edged blade 33ins. £110

A fine early 17th Century German semi swept hilted broadsword. £140

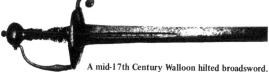

A mid-17th Century Walloon hilted broadsword, straight double edged blade, 36½ins. £250

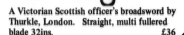

A Victorian Scottish officer's broadsword by
Thurkle, London.   Straight, multi fullered
blade 32ins.   £36

A Scottish military issue broadsword. The plain
blade 31½ins. with central fuller.   £55

A good, late 18th Century Scottish officer's broadsword. The double edged blade has fuller at forte,
with Running Wolf mark and traces of 'Andria' in
fullers.   £105

A mid-18th Century Scottish basket hilted horse-
man's broadsword.   The double edged straight
fullered blade 32½ins.   £125

A good fine early 18th Century Military Scottish
broadsword, broad straight double edged blade,
34¾ins.   £175

A good, silver mounted Bade Bade, recurving blade 7½ins. With a tulip hilt covered with filigree and embossed Eastern silver with foliate patterns. £32

A good, silver mounted Bade Bade, recurving blade 8½ins. Elaborately carved hilt with floral patterns of tulip form.   £32

An Acheen Bade Bade inlaid with brass inscriptions on both sides, curved blade 9½ins.   £32

A fine, silver overlaid, tulip hilted Bade Bade, blade 6ins.   The entire hilt and sheath covered with repousse, foliate decorated silver.   £34

A good Malayan Bade Bade, blade 10ins.   £36

11

An Arab Jambiya, blade 8ins. with central rib. The horn hilt with
white metal rivets and steel stud decoration. In its 'U' shaped sheath,
decorated with silver. £25

A fine quality Caucasian Jambiya, curved blade 7ins. with central rib.
The entire hilt and sheath of silver, the front nielloed overall with
scrolled foliage and with small overlaid gold plaques. £36

A good Arab Jambiya, blade 7ins. with central rib. With horn hilt overlaid
with silver filigree roundels, the back covered with silvered sheet. £46

A fine silver gilt Balkan Jambiya, curved blade 12½ins. with central rib. The
blade decorated with gold damascus at forte and the hilt covered with plain
Eastern silver gilt sheet. £95

An early 18th Century Rajput dagger Katar, bi fullered watered blade 9ins. £18

A heavy all metal Indian Katar, tapering blade 11ins. Blade with reinforced point has fluted decoration at forte. £22

A fine, Indian gold damascus Katar, plain blade 12ins. With the hilt covered in gold damascus foliate patterns and baluster grips. £36

An Indian scissors Katar, central blade 7ins., outer blade 7½ins. The guard of scrolled brass surmounted by peacocks. £48

A fine Indian Katar, blade 10½ins. Blade has large armour piercing point. With a steel hilt with some gold damascus patterns. £70

A Cossack dagger Kindjal, single narrow fullered blade 17ins. £22

A good, Cossack silver mounted Kindjal, broad double edged blade 18½ins. Blade has deep central fullers and is struck with two marks at forte.   £48

A silver mounted Cossack Kindjal, double edged blade 13ins. Blade with pronounced point.   £60

A decorative Caucasian Kindjal, blade 17ins., with central fuller. £90

A fine, Caucasian Kindjal, double edged blade 13ins. Blade has deep central fuller with gold damascus decoration of animals, foliage, etc. With bone grips inlaid in gold damascus studs.   £100

A good, Russian dagger Kindjal, double edged swallow diamond sectioned blade 10ins. With iron hilt and sheath scroll inlaid with arabesque silver with pellet borders.   £130

A silver mounted Lombok Kris, plain straight blade 12ins. £22

A large Bali Kris, wavy watered steel blade 17½ins. £35

A large Bali Kris, straight watered blade 18½ins. £50

A good, Malay kingfisher Kris, straight blade 12ins. £70

A large, decorative Balinese Kris, wavy watered blade 17½ins. £110

A fine and rare Bali Kris, blade 14ins. having entirely different watering on either side. £175

A good, heavy, Tibetan exorcising dagger Phurbu, 10ins. long. With a triblade of iron emanating from Makara mask, a brass body and three Bodhisatva heads with vajra terminal. £15

A good, Indian Bhuj, blade 5½ins. with clipped-back tip. With Cont. screw in dagger at base, blade 4ins. complete with wooden blade sheath. £40

A fine quality Indian dagger Kard. With a fine blade, 9½ins. long with watered steel chevron patterns and a reinforced point, also with a steel hilt with some gold damas foliate and other patterns. £80

A pale green jade hilted Indian dagger Khanjar. The double edged, slightly curved blade 9½ins. and the hilt carved with foliage in shallow relief overall. £140

A good old Persian dagger. The spear shaped blade 9¼ins., pierced with openwork panels with a reinforced point, decorated overall in gold damas decoration and inscriptions. £250

An Indian Khanjar. The blade, 10½ins. long, is of watered steel and has a reinforced point. With a green hilt of mutton fat jade decorated with some foliate patterns. £320

A Japanese cloisonne dagger, straight, multi fullered double edged blade 6 ins. Hilt and sheath covered in cloisonne floral patterns with similar rectangular belt mount. £38

An attractive Japanese dagger Tanto, blade 10¾ins. With habaki copper tsuba and kodzuka with silver inlaid handle. Inlaid overall with silver and gold. £55

A good Japanese carved bone dagger 14½ins., blade 8ins., nicely carved overall. £70

A Japanese dagger Aikuchi, slightly curved blade 11½ins. With brass nanako, small oval tsuba, tang pierced with three mekugi ana and plain kashira of wood. £100

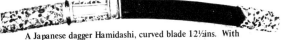

A Japanese dagger Hamidashi, curved blade 12½ins. With small oval copper tsuba and tang with single mekugi ana. The intricately chiselled hilt is of solid copper with traces of silver plating and gilt. £100

A Japanese dagger Aikuchi, straight heavy blade 7½ins. with pronounced clipped-back edge. With a cord bound hilt. In its lacquered wooden saya covered in imitation birch bark with kodzuka, blade 3¼ins. £110

A good, Nazi S.A. dagger. The blade bears RZM mark and
also marked 'M7/83'. With plated mounts.   £36

A Nazi Luftwaffe Presentation bayonet by P.D. Luneschloss.
Etched with Luftwaffe eagle and inscription 'Erinnerung an
Meine Dienstzeit'. With plated mounts, in its metal scabbard. £48

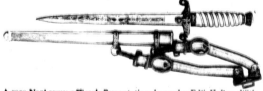

A rare Nazi army officer's Presentation dagger by F.W. Holler. With
etchings on both sides of Nazi eagle and foliage patterns. With engraved
crossguard and yellow grip. In its plated metal sheath.   £95

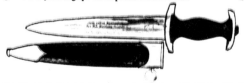

A rare, Nazi S.A. Presentation dagger by Plucker. The blade etched
with presentation inscription, 'Zum ersten Spatenstich der S.A.
Siedlung Krefeld S.A. Standarte 40'.   £105

A very rare Nazi Teno Officer's dagger, blade 11ins. by Eickhorn. £150

An interesting Georgian naval officer's dirk, blade 15½ in. With
oval solid steel guard with fluted edge, plain steel mounts and a fluted
ivory grip   £22

A Georgian naval officer's dirk, circa 1820. Tapering blade 6 in. of
flattened diamond section, frost etched in military trophies and
foliage. Small oval copper gilt guard with scale decoration   £27

A Georgian dirk, blade 8½ in. With white metal quillons with ball
terminals, white metal mounts, diced black grip. In its white metal
sheath   £46

A good, Georgian naval officer's dirk. Tapering blade 6 in. of flattened
diamond section with blued and gilt etched decoration of foliage. With
copper gilt guard of cruciform shape and a turned ivory hilt. In its
copper gilt metal mounted leather sheath   £60

A fine Victorian naval officer's dirk, single edged blade 18 in. etched
with crowned fouled anchor and 'V.R.' cypher amidst profuse scrolls
                                                                                  £105

A fine Nazi Naval Officer's presentation dirk, the blade blued and gilt
with Nazi Eagle and fouled anchor   £190

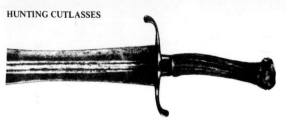

A large, 18th Century hunting sidearm, broad spatulate blade 20ins.
Blade with double fullers. With reversed, plain steel crossguard and
a staghorn hilt.   £11

A late 19th century Imperial German prize hunting cutlass, blade 15in.
With a brass shellguard, a reversed hoof crosspiece and a brass mounted
staghorn hilt mounted with three brass acorns. In its brass mounted
leather sheath.   £50

A 19th century German Presentation hunting cutlass, 16in. Bears various
etchings. With a brass shellguard, reversed hoof quillons, brass mounts and
a staghorn hilt mounted with three brass acorns.   £55

A Georgian figure of 8 naval cutlass, circa 1800, single edged blade 29ins.
Blade stamped at forte. With a black figure of 8 sheet metal guard and a
ribbed iron grip with old white paint.   £60

An early 18th Century hunting hanger, curved blade 21ins. With an elaborately pierced shellguard. £45

A good, mid-18th Century German hunting hanger, curved clipped-back blade 22ins. long, and with small brass shellguard. £54

A mid-18th Century French silver mounted hunting hanger, slightly curved blade 18½ins. With reversed quillons of silver, a spiral wooden hilt and a fluted pommel. £55

A good, mid-18th Century silver mounted hunting hanger, plain curved blade 23ins. With a reversed silver crossguard with reeded decoration. £70

A Georgian, military naval type hanger, slightly curved blade 24ins. With sheet metal double shellguard, knucklebow and mounts and a bone grip. £80

HUNTING SWORDS

A good, mid-18th Century European hunting sword, plain straight single edged blade 21½ins. with back fuller.   £48

A fine George III silver hilted hunting sword, plain straight edged multi-fullered blade 22ins.   With a silver shellguard, silver knucklebow and mounts, a silver pommel in the form of a mask of an old bearded man and a fluted black grip.
                                                                                                    £70

A fine, French 18th Century silver mounted hunting sword, curved blade 23ins. Blade etched with decorations and an inscription.   With a silver crossguard with an embossed central panel and a silver wire bound ivory hilt.   £95

An early 17th Century Swiss hunting sword, straight single edged T sectioned blade 34ins.   The back of the blade waved by alternate scalloping of edges. £120

One of a pair of early 19th century hunting hangers, bifullered double edged blades 20ins.   With cruciform hilts, etched single shellguards and turned ivory grips.   In their steel mounted brown leather scabbards, with belt hooks and etched mounts.   £145

A Japanese Katana, blade 26ins., signed on both sides 'Showa 18th year' (1943), in its black lacquered wooden saya, partly bound with cord. £155

A Japanese sword Katana, blade 27ins., in shirasaya attributed to Echizen Hirotaka, broad irregular choji hamon, mokume hada, two mekugi ana. £250

A Japanese, World Wars One and Two sword Katana, blade 21ins., shinogi zukuri, medium straight hamon, tight mokume hada, in shin-gunto mounts with leather-covered scabbard. £470

A good, Japanese sword Katana, blade shinogi zukuri 30½ins., straight close grain tang with three mekugi ana, chiselled iron tsuba, some gold and silver detail en suite with fuchi kashira. £960

An attractive and most desirable Japanese sword Katana, blade 29½ins., shinogi-zukuri, tight ayasugi hada, small regular choji hamon, long tang and point, signed 'Tairusai Munehird Korewosaku', dated Genji 1st year. £1,450

A good old bowie type knife, blade 10ins., stamped 'V.R.' G. Beardshaw. £30

A good Victorian bowie knife, pronounced clipped-knife blade 7¾ins., in its original leather covered sheath. £75

A fine, large Victorian bowie knife by 'Thos. Ibbotson & Co., Makers, Sheffield'. Broad pronounced clipped-back blade 11½ins. With white metal scrolled crossguard embossed in foliage, plain horn grips and pommel of two white metal panels. £185

A very fine Victorian bowie-type folding knife. The clipped-back blade stamped 'Of the best Quality' and with the maker's name, 'Snow, Portsea' is 5ins. in length. With a German silver mounted hilt with acanthus leaf crosspiece and a horse's head pommel. £280

A large Victorian bowie knife of classic proportions, 17ins. overall. The heavy clipped-back blade, stamped 'Alfred Hunter' at forte is 11ins. and the back of this is over ¼in. in thickness and therefore is obviously intended as a cutting weapon. £310

An unusual Ceylonese knife Piha Kaetta, blade 7½ins. Blade overlaid with scroll engraved brass at forte. Silver grip of scroll decorated in repousse. £25

A set of Persian knives. Comprising a Kard, blade 6ins. with finely watered pattern, with a white stone hilt with gold damascus and iron ferrule. Four small knives, blades 2½ins. to 3½ins., one with a silver hilt. £32

A good Burmese knife D.H.A., single edged blade 9½ins. £38

An Indian Knife of Ayda Katti type, broad blade 11½ins. With a wooden hilt of rectangular section almost entirely covered with engraved brass. £46

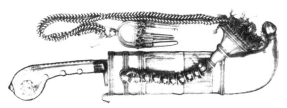

A fine Coorg silver mounted Pichangatti knife, blade 7ins. With an Eastern silver hilt. £80

KNIVES

A South American silver mounted gaucho knife, blade 8ins. With embossed silver hilt. In its embossed silver sheath.   £23

A South American gaucho silver mounted knife, by F. Herder, blade 8½ins. With octagonal grip of silver. In its silver mounted leather sheath.   £34

A silver mounted South American gaucho knife, blade 7¼ins. With turned hilt of silver. In its solid silver sheath, finely chiselled in birds, flowers, etc. £40

RAPIERS

A good, Spanish cup-hilt rapier, circa 1700, slim blade 38ins. of diamond section.   £190

A fine Italian 17th Century swept hilt rapier, tapering double edged blade 40ins.   £250

A mid-17th Century gunner's stiletto, triangular sectioned blade 9½ins. to turned pedestal. With steel crosspiece with spiral twisted terminals and spiral carved horn grip.   £70

A 17th Century Italian stiletto (stylet), triangular blade 5¼ins. With cast brass hilt with turned quillons, the hilt in the form of monkeys.   £70

TRENCH DAGGERS

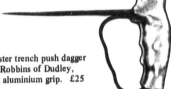

A scarce World War I knuckleduster trench push dagger of the pattern manufactured by Robbins of Dudley, fluted blade 4½ins. With shaped aluminium grip. £25

A scarce World War I, U.S. trench knife.   £30

A World War I trench fighting sword, broad leaf shaped blade 18 ins. With folding guard and cord bound hilt.
£100

A good, Georgian Cavalry officer's sabre,
circa 1800, curved blade 29ins.   £46

A Georgian Cavalry officer's sabre, curved
blade 30½ins.   £75

A rare, French revolutionary-period officer's
sabre, circa 1795, curved blade 33ins.   £130

A very fine, Georgian officer's Presentation sabre, curved blade 32ins. The
finely blued and gilt blade bears etched decoration over its entire length,
including the crown.
                                     £600

A rare, Georgian volunteer officer's Spadroon, circa 1780, plain straight single edged blade 31 ins. With copper guard, intricately interlaced plain knucklebow and fluted ivory grip. £36

A good, Georgian officer's '5-ball' hilt Spadroon, circa 1800, straight single edged blade 29½ins. With knucklebow and mounts and sideloop incorporating '5-ball' design. £36

A Georgian officer's Spadroon, circa 1780, plain straight single edged blade 32ins. £42

A good, Georgian officer's '5 ball' hilt Spadroon, circa 1780, straight single edged blade 32ins. £127

A curious Burmese sword, blade 18in. The hilt of horn carved in the form of a bird's head. Contained in its carved wooden scabbard imitating the plumage and body of a bird. £15

An Indian sword Firangi, early 18th Century, straight bi-fullered blade 34ins. With a watered steel hilt of classic form. £28

A fine, old Chinese double sword, blades 16½ins. With brass crossguards demon masks, rounded wooden grips and pommels with traditional engraved scrollwork. £38

A fine, Chinese sword, straight double edged blade 27ins. With plain heavy German silver crossguard and pommel of traditional design, grip inlaid in brass wire. In its wooden scabbard with large, plain German silver mounts, and some brass wire inlay to hilt. £65

A silver mounted Burmese dha, slightly curved blade 25½ins. The blade inlaid in silver on both sides and the hilt covered in sheet silver. In its wooden scabbard covered in low grade Eastern silver sheet and brass decoration. £70

A good Indian shamshir hilt sword, slightly curved blade 28ins. Blade with clipped-back point. With bird's head pommel and knucklebow. £85

An Indo-Persian shamshir, curved blade 30ins. With solid copper gilt guard and leopard's head pommel. In its solid copper gilt scabbard. £100

A large Ceremonial Executioner's Tulwar, broad curved blade 24ins. The blade with a reinforced tip inlaid with brass designs. With an old iron hilt with straight crosspiece. £160

A good Turkish sabre Kilij, curved broad clipped-back blade 27½ins. The blade inlaid in gold and the hilt and scabbard mounted with Turkish silver. With two piece carved horn grips and a black leather pommel carved with leaves. £230

A Prussian 1889-pattern Infantry Officer's sword, blade 27½ins. £38

A Nazi Luftwaffe officer's sword, blade 28½ ins. by Eickhorn.   £45

A fine .mperial German Cavalry officer's sword, slightly curved partly
fullered blade 32ins., by Neumann, Berlin.   £55

An Imperial German Naval Officer's sword, clipped-back blade 29ins. £63

A Presentation Imperial German student's duelling Schlager, straight blade 33ins. With large basket guard and sharkskin covered grip. In its metal plated scabbard.   £75

A rare, Nazi Police officer's Presentation Boy Service sword, 32½in. by Weyersberg.   £80

A Nazi Naval Officer's sword, plain blade 27ins.   £130

A good, Nazi Naval Officer's sword, slightly curved blade 29ins. £185

33

A 1796-pattern Georgian officer's sword. Engraved on the backstrap 'J.J. Runkel Solingen'. £24

A William IV 1822-pattern field officer's sword, slightly curved pipeback, clipped-backed blade 32ins. by Moore 'Late Bicknell & Moore', London. £27

A good, Georgian 1803-pattern officer's sword, curved blade 32ins. £28

A good, early Victorian 1822-pattern Infantry officer's sword, slightly curved blade 32½ins. £30

A good, Victorian 1822-pattern Infantry officer's sword, slightly curved blade 31½in., by 'Pillin & Son, Manufacturer, London'. £30

A Victorian 1822-pattern Cavalry officer's sword of the 18th Hussars, slightly curved blade 35ins., by Henry Wilkinson. £40

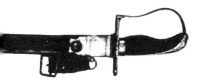

A good, Georgian naval officer's sword, single edged blade 23ins. £42

A rare, Scandinavian early 19th century Cavalry sword, straight plain single edged blade 33in. £42

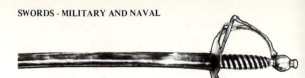

A brass-hilted Infantry hanger, circa 1750, plain curved blade 24½ins., with ordnance stamp, heart-shaped guard, two loops, spiral grip, ovoid pommel.   £58

A Victorian 1822-pattern Infantry officer's sword by Rogers. Slightly curved blade 32ins. Bears etchings and presentation inscription.   £75

A rare, Georgian 1796-pattern Infantry officer's sword of the Royal Regime, straight single edged blade 32ins., engraved 'J J Runkel Solingen'.   £

A good, 1796-pattern Infantry officer's sword, straight single edged blade, 32ins.   £100

A basket hilted Dragoon sword, circa 1775. The straight, single edged, fullered blade 32ins. long. The basket fenestrated and fluted, with six facetted recurved bars.   £115

A Victorian officer's 1831-pattern mameluke sword by Thurkle, London. The slightly curved, clipped-back blade, 32½ins.   £120

A fine, Georgian 1822-pattern general officer's sword, clipped-back blade, 33ins.   £150

rare, 1796-pattern Household Cavalry trooper's sword. The plain, straight, ngle edged blade 34ins. Stamped on backstrap 'Josh H. Reddel'.   £210

A late 18th Century Cont.smallsword. The slim , tapering blade 31½ ins. £3

A Georgian cut steel hilted smallsword, circa 1800. The plain , triangular, tapering blade 32ins. long of hollow ground section. With a cut steel pierced oval shellguard. £5

A good, Georgian mourning smallsword. The blade 32½ins. of hollow triangular section. With a steel hilt with pierced and engraved oval guard. £74

An attractive mid-18th Century German smallsword. The colichemarde blade 32½ins., bears various etchings. With an intricately pierced steel hilt. £1

A fine mid- 18th Century Cont,hallmarked silver hilted smallsword. The silver hilted blade 32¾ins. With a hallmarked silver hilt and a silver wire and tape bound grip. £250

A United States society sword by Luker, plated blade 27ins.   £19

A United States society sword, straight blade 18in.   With elaborate
pierced cruciform crossguard, chain guard, close helmet pommel and
ivory grip engraved with symbols.   £26

A good, United States society sword by Horstmann of Philadelphia. £36

A good, United States society sword by Parson & Co. of St. Louis.   £36

An interesting Victorian naval officer's Presentation sword of United
States interest, slightly curved , piped-back, clipped-back, Spanish
made blade 29ins.   £100

An attractive Japanese Wakizashi, blade 17¾ins. Gunome hamon blade, with two mekugi ana, mumei but shin shinto and silver neko gake hamabki. £75

A Japanese Wakizashi, slightly curved blade 20ins. With an iron quatrefoil tsuba and fuchi kashira with raised designs in silver and a pierced tang with two mekugi ana. £150

An attractive Japanese Wakizashi, blade 19¾ins. Blade signed, 'Yamato Nokami Yoshimichi', with two mekugi ana, shallow tori, gunome hamon with nie clusters, ayasugi and mokume hada. £260

An attractive Japanese short sword Wakizashi, blade 13¼ins. Unobi-zukuri blade signed 'Iganokami Fujiwara Kinmichi Kikumon', with one mekugi ana, intensively active overall with mokume and ayasugi hada. £400

A fine, late 18th Century Wakizashi for presentation to a temple, blade 19ins. Blade hira zukuri, carved with horimono of vajra hilted ken, bonji and ken tailed dragon protecting sacred jewel, with a broad hamon and signed, 'Shirakawa No Kashin Tegarayama Masashige', dated 1789. £1,100

An unusual, French or Italian flintlock blunderbuss pistol, 12½ins., half octagonal barrel 6½ins. Fullstocked, with engraved rounded lock and steel furniture.   £200

A charming and unusual, late 18th Century French steel barrelled flintlock blunderbuss pistol, 7¾ins., swamped barrel 3¾ins. £290

A good, brass barrelled flintlock blunderbuss pistol, circa 1800, 9ins., bell mouthed barrel 4½ins. London proved, with plain fullstock and slightly flattened rounded butt.   £360

A rare, late 17th century brass barrelled flintlock blunderbuss pistol, 13½ins., slightly swamped barrel 7¼in. London proved Fullstocked, with swan neck cock, brass mounts.   £390

BLUNDERBUSS

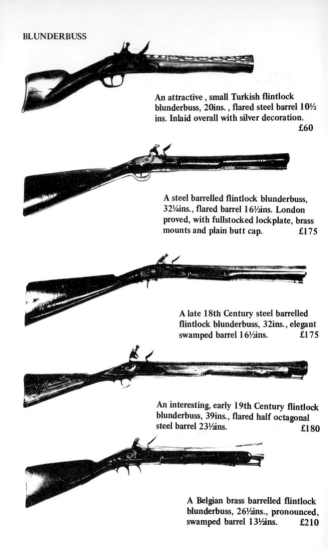

An attractive, small Turkish flintlock blunderbuss, 20ins., flared steel barrel 10½ ins. Inlaid overall with silver decoration.
£60

A steel barrelled flintlock blunderbuss, 32¼ins., flared barrel 16½ins. London proved, with fullstocked lockplate, brass mounts and plain butt cap.
£175

A late 18th Century steel barrelled flintlock blunderbuss, 32ins., elegant swamped barrel 16½ins.
£175

An interesting, early 19th Century flintlock blunderbuss, 39ins., flared half octagonal steel barrel 23½ins.
£180

A Belgian brass barrelled flintlock blunderbuss, 26½ins., pronounced, swamped barrel 13½ins.
£210

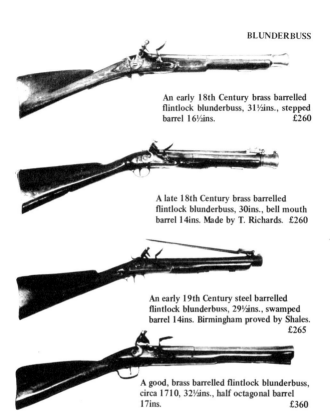

An early 18th Century brass barrelled flintlock blunderbuss, 31½ins., stepped barrel 16½ins. £260

A late 18th Century brass barrelled flintlock blunderbuss, 30ins., bell mouth barrel 14ins. Made by T. Richards. £260

An early 19th Century steel barrelled flintlock blunderbuss, 29½ins., swamped barrel 14ins. Birmingham proved by Shales. £265

A good, brass barrelled flintlock blunderbuss, circa 1710, 32½ins., half octagonal barrel 17ins. £360

A good, brass barrelled flintlock blunderbuss with spring bayonet, 27½ins., sweaped barrel 12ins. Tower proved, fullstocked with bayonet released on top of barrel by sliding top thumb catch. £390

A fine, 11-bore flintlock Yeomanry carbine, 44ins., barrel 27½ins.
Birmingham proved with slightly rounded lockplate engraved
'Raper, Leeds'. Fullstocked, with regulation brass mounts, a steel
ramrod and original triangular socket bayonet, 16ins., by S. Hill. £290

A rare, .66 in. Paget flintlock Cavalry carbine, 31ins., London and military
proved sighted barrel 16ins. Fullstocked, with stepped lockplate, a sliding
safety bolt to throathole cock, raised pan, regulation brass mounts, side
saddle bar with ring and swivel ramrod.   £410

A good, rare, 16-bore Elliot's-pattern volunteer Cavalry flintlock carbine,
circa 1800, 44ins., barrel 28ins. With Tower private proofs, plain flatlock
and swan neck cock with double line engraved borders and impressed
'Mather, Newcastle'. Also with nicely figured walnut fullstock with standard
pattern military brass mounts, steel saddle bar and ring on left of stock with
its original triangular socket bayonet by Wooley & Deakin. In its brass
mounted leather scabbard.   £450

A pair of 22-bore flintlock duelling pistols, 16in., octagonal barrels 10¼in. French cocks, rainproof pans, rollers on frizzen springs and capstan set screw triggers.   £400

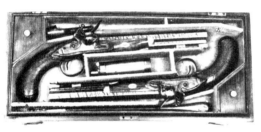

A pair of 34-bore flintlock duelling pistols by W. Smith, 15¾ins., heavy octagonal twist barrels 10ins. Halfstocked, with rainproof pans, high fences, rollers on frizzen springs and capstan set screws.   £650

A good quality pair of 34-bore flintlock duelling pistols by Mortimer, 16ins., octagonal twist barrels 10¼ins. Rainproof pans, rollers on frizzen springs and silver lined vents.   £675

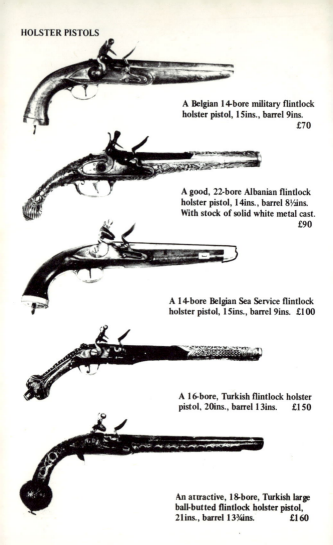

A Belgian 14-bore military flintlock holster pistol, 15ins., barrel 9ins.
£70

A good, 22-bore Albanian flintlock holster pistol, 14ins., barrel 8½ins. With stock of solid white metal cast.
£90

A 14-bore Belgian Sea Service flintlock holster pistol, 15ins., barrel 9ins. £100

A 16-bore, Turkish flintlock holster pistol, 20ins., barrel 13ins. £150

An attractive, 18-bore, Turkish large ball-butted flintlock holster pistol, 21ins., barrel 13¾ins. £160

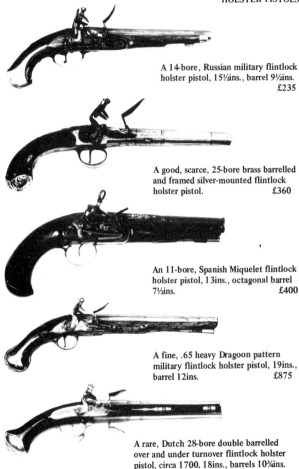

A 14-bore, Russian military flintlock
holster pistol, 15½ins., barrel 9½ins.
£235

A good, scarce, 25-bore brass barrelled
and framed silver-mounted flintlock
holster pistol.   £360

An 11-bore, Spanish Miquelet flintlock
holster pistol, 13ins., octagonal barrel
7½ins.   £400

A fine, .65 heavy Dragoon pattern
military flintlock holster pistol, 19ins.,
barrel 12ins.   £875

A rare, Dutch 28-bore double barrelled
over and under turnover flintlock holster
pistol, circa 1700, 18ins., barrels 10¾ins.
Fullstocked, with rammer on one side and
plain trigger guard.   £1,000

A charming flintlock boxlock muff pistol, circa 1820, 4¼ins., screw-off barrel. London proved. £115

A charming flintlock boxlock muff pistol by Grierson, circa 1800, 4¾ins., turn-off barrel 1½ins. London proved. £115

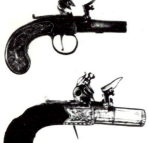

A charming flintlock boxlock muff pistol by Brasher of London, 4¾ins., screw-off barrel. London proved. £140

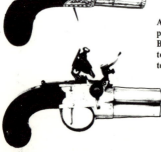

A charming flintlock boxlock muff pistol, 4¾ins., turn-off barrel 1in., Birmingham proved. With sliding top thumb safety catch, locking tension sprung teardrop frizzen. £175

A good quality, brass barrelled brass framed flintlock boxlock over and under muff pistol, 5½ins., barrels 1¾ins. £225

A 24-bore Cont. military flintlock musket (probably Austrian), 49ins., barrel 33ins. Fullstocked, with two plain steel barrel bands retained by spring catches and steel mounts. Lockplate dated '1856'. £95

A 10-bore Victorian Brown Bess flintlock musket, 55ins., barrel 39ins. Military proved and engraved '42 Reg' (Royal Highlanders). Fullstocked with flattened lockplate, regulation brass mounts, steel ramrod and sling swivels. The stock deeply impressed with government sale marks and 'Barnett Gunmakers London'. £230

A rare, 10-bore .42 in. Brown Bess flintlock musket, 58ins., barrel 42ins. Tower proved and engraved '86'. Fullstocked, with linear engraved lockplate with 'Ketland & Co.', regulation brass mounts, an oval escutcheon, steel ramrod and sling swivels. The stock carved with 'Y + C 133'. £240

An 18th Century cannon barrelled flintlock boxlock pistol 7½ins., screw-off barrel 2ins., by Clarkson, London. **£140**

A scarce French 14-bore model Military flintlock Cavalry pistol 14ins., barrel 7¾ins., dated 1808 at breech. **£175**

A mid-18th Century Queen Anne-type 20-bore cannon barrelled flintlock pistol, by Stanton, London, 12ins., screw-off barrel 5½ins. **£180**

A scarce, French double barrelled side by side flintlock boxlock cannon barrelled overcoat pistol, circa 1800, 8¾ins., turn-off cannon barrels 3¼ins. **£230**

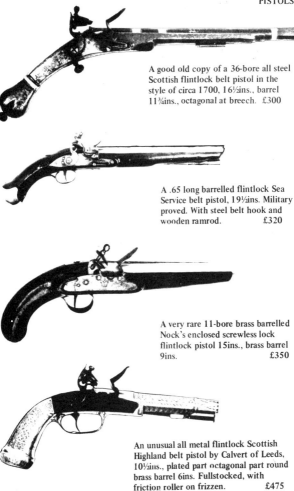

A good old copy of a 36-bore all steel Scottish flintlock belt pistol in the style of circa 1700, 16½ins., barrel 11¾ins., octagonal at breech. £300

A .65 long barrelled flintlock Sea Service belt pistol, 19½ins. Military proved. With steel belt hook and wooden ramrod. £320

A very rare 11-bore brass barrelled Nock's enclosed screwless lock flintlock pistol 15ins., brass barrel 9ins. £350

An unusual all metal flintlock Scottish Highland belt pistol by Calvert of Leeds, 10½ins., plated part octagonal part round brass barrel 6ins. Fullstocked, with friction roller on frizzen. £475

51

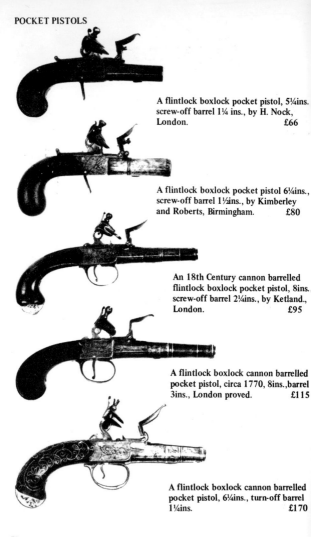

A flintlock boxlock pocket pistol, 5¼ins. screw-off barrel 1¼ ins., by H. Nock, London. £66

A flintlock boxlock pocket pistol 6¼ins., screw-off barrel 1½ins., by Kimberley and Roberts, Birmingham. £80

An 18th Century cannon barrelled flintlock boxlock pocket pistol, 8ins. screw-off barrel 2¼ins., by Ketland., London. £95

A flintlock boxlock cannon barrelled pocket pistol, circa 1770, 8ins., barrel 3ins., London proved. £115

A flintlock boxlock cannon barrelled pocket pistol, 6¼ins., turn-off barrel 1¼ins. £170

A flintlock boxlock pocket pistol, 6ins.,
turn-off browned twist barrel 1½ins.,
Birmingham proved and engraved
'H. Nock, London'.   £185

A good, French late 18th Century
flintlock boxlock pocket pistol,
7¼ins., barrel 2¼ins.   £200

A good, brass framed double barrelled
over and under tap action flintlock
boxlock pocket pistol, 6ins., screw-off
barrels 1¼ins., by Williamson, Hull.
                                     £225

One of a small pair of flintlock boxlock
pocket pistols by Laugher of London,
4¾ins., turn-off barrels 1¼ins. London
proved.   £310

One of a fine quality pair of French
cannon barrelled boxlock flintlock
pocket pistols by Bizovard, circa 1780
6ins., turn-off barrels 1½ins.   £500

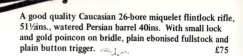

A good quality Caucasian 26-bore miquelet flintlock rifle, 51½ins., watered Persian barrel 40ins. With small lock and gold poincon on bridle, plain ebonised fullstock and plain button trigger. £75

A good North African 16-bore miquelet flintlock gun, 72½ins., octagonal barrel 57ins. With large, curved cock, the cock and frizzen scroll engraved and hinged dog catch on lock plate. £120

A good 22-bore flintlock sporting gun by Bate, circa 1760, 55ins., barrel 38½ins. With silver foresight stepped lockplate, roller on frizzen spring and rounded skeleton butt. £260

A .65 Elliot's Pattern military flintlock carbine, 43½ins., barrel 28½ins. Tower proved. With fullstocked lockplate, regulation brass mounts, slotted steel sling swivels. £370

A fine and rare early 17th century, French or Italian snaphaunce sporting gun, 70½in., part octagonal blued barrel 56¼in. Fullstocked in walnut, with brass sights and channel rear sight. £600

A single barrelled 16-bore flintlock sporting gun by John Manton, 47½ins., twist barrel 32ins. £240

A scarce, mid-18th century Danish 16-bore flintlock sporting gun, 54ins., good quality Persian barrel 38½ins. £270

A good, rare, 16-bore breech loading flintlock sporting rifle by Waters, circa 1775, 50½ins., swamped barrel 34ins. London proved, the breech with vertical loading chamber. £550

A fine quality German 20-bore flintlock sporting rifle, circa 1730, 55½ins., half octagonal barrel 40½ins. £685

A fine quality, French double barrelled flintlock sporting gun by Barbet of Paris, 48½ins., blued barrels 32ins. Paris proved. Barrels extensively gilt at breech with thunderbursts around proof marks. £775

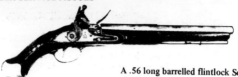

A .56 long barrelled flintlock Sea Service belt pistol 19½ins. barrel 12ins.   £140

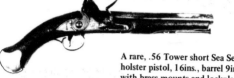

A rare, .56 Tower short Sea Service flintlock holster pistol, 16ins., barrel 9ins. Fullstocked, with brass mounts and lockplate marked with 'Tower' crown.   £170

A .65 long barrelled flintlock Sea Service belt pistol, 19½ins., barrel 12ins.   £210

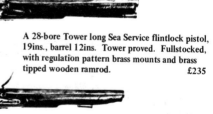

A 28-bore Tower long Sea Service flintlock pistol, 19ins., barrel 12ins. Tower proved. Fullstocked, with regulation pattern brass mounts and brass tipped wooden ramrod.   £235

A good .56 long barrelled flintlock Sea Service belt pistol, 19½ins., barrel 12ins.   £250

A .65 long barrelled East India Company flintlock Sea Service belt pistol, 19½ins., barrel 11ins. £270

A rare, .56 William IV short Sea Service flintlock holster pistol, 15ins., barrel 9ins. Fullstocked, with brass mounts, belt hook and a swivel ramrod. £280

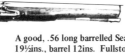

A good, .56 long barrelled Sea Service belt pistol, 19½ins., barrel 12ins. Fullstocked, with military proofs, regulation brass mounts, steel belt hook and brass tipped wooden ramrod. £310

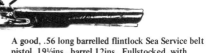

A good, .56 long barrelled flintlock Sea Service belt pistol, 19½ins., barrel 12ins. Fullstocked, with military proofs, steel belt hook and a brass tipped wooden ramrod. £325

A .56 long barrelled flintlock Sea Service belt pistol. 19½ins., barrel 12ins. Fullstocked, with regulation brass mounts, steel belt hook and a brass tipped wooden ramrod. £350

57

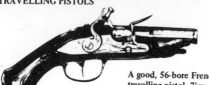

A good, 56-bore French or Italian flintlock travelling pistol, 7ins., half octagonal barrel 3½ins. With fullstocked scroll engraved lock and engraved steel furniture stock. £105

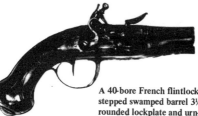

A 40-bore French flintlock travelling pistol, 7ins., stepped swamped barrel 3¼ins. With a fullstocked rounded lockplate and urn-finialled trigger guard. £130

A good flintlock boxlock travelling pistol by Anderson, 7ins., barrel 2¼ins. Fitted with sidespring bayonet, blade 1½ins. London proved, with floral engraved sideplates and a folding trigger. £135

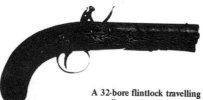

A 32-bore flintlock travelling pistol, circa 1810, 9½ins. overall, round twist barrel 5ins., with flat top engraved 'London'. £185

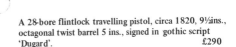

A 28-bore flintlock travelling pistol, circa 1820, 9½ins., octagonal twist barrel 5 ins., signed in gothic script 'Dugard'. £290

An early 18th century Queen Anne cannon barrelled flintlock sidelock travelling pistol, 12ins., barrel 5ins., by I. Smith, London proofs. £310

A good quality flintlock boxlock sidelock travelling pistol by Nock, 8½ins., turn-off barrel 3ins. London proved with bolted cock, shaped frizzen spring under pan and a walnut butt. £430

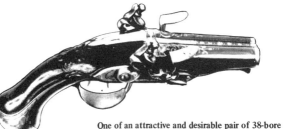

One of an attractive and desirable pair of 38-bore double barrelled French flintlock travelling pistols, circa 1780, 7¾ins., barrels 3½ins. With fullstocked border engraved locks. £1,030

59

A good and scarce, .36in pump up all steel air rifle, 43ins., barrel 24¼ins. With brass rifled barrel liner, a rear loading cavity with lever, an unusual skeletonised open butt and concealed button trigger. Black lacquered overall. £90

A good , .36ins pump up walking stick air gun, 38½ins., brass rifled barrel 21ins. With button trigger, ivory handle and muzzle plug screw on cap. Black lacquered overall. £140

A rare, 12.8mm., Austrian repeating air rifle model 1799, 34¾in., barrel 18½in. With plain fullstocked forend with brass ramrod pipes, brass trigger guard and left side mechanism plate, the steel lock plate with cocking hammer and maker's name engraved. £155

An interesting 70-bore walking stick air gun, 37½ins overall. With a screw on handle of wood concealing the top of the pump. £199

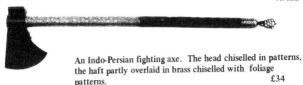

An Indo-Persian fighting axe. The head chiselled in patterns,
the haft partly overlaid in brass chiselled with foliage
patterns.   £34

A fine, silver damascened Indian axe from the Hindu Kutch,
22ins., crescent head 5¼ins. Silver damascened overall,
terminating in a squared spike. With a steel handle, partly
damascened, which unscrews to reveal a concealed dagger,
blade 9ins., terminating in a lotus finial.   £42

A fine, old North American Indian tomahawk pipe, head 22½ins.,
height 9¼ins., iron blade 4¼ins. With brass top with well
executed incised borders and chevron decoration to bowl.
Probably made for trade exhibition display.   £105

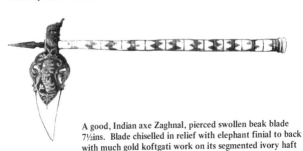

A good, Indian axe Zaghnal, pierced swollen beak blade
7½ins. Blade chiselled in relief with elephant finial to back
with much gold koftgati work on its segmented ivory haft
with polychrome incised geometric decoration.   £115

A Cromwellian trooper's half armour, lobster tail helmet, fluted skull and four piece articulated neck lames.   £180

A Cromwellian suit of Pikeman's armour. £285

A Japanese half suit of armour bearing a gilt monogram. £375

Mid 17th century cuirassier's armour, with closed burgonet with roped false comb and gorget rim. £420

A Japanese half suit of armour in good condition with fluted Kabuto, four neck lames, plain mempo and eight plate lames. £450

A fine complete suit of Pikeman's armour, English circa 1600, comprising 'Pot' gorget breast, backplate and tassets.   £450

A good Japanese half suit of armour complete with arm and shoulder defences. £475

A complete suit of Japanese armour with black lacquered laced body plates. £550

A well constructed late Victorian suit of armour made in the 16th century style, 6ft 1ins high. £700

16th century half suit of black and white Landsknecht. £750

A complete set of etched armour with matching shield. £1,300

A magnificent antique suit of armour in the Gothic style, manufactured for the 16th Earl of Shrewsbury, circa 1840. £5,000

Officer's badge of
The Tyneside Scottish,
1918   £15

A Victorian Officer's
silver plated 1855 -
pattern shako badge
of The Second West
York Militia   £21

Officer's badge of The
East Lancashire Regi-
ment, pre 1881   £22

Cap badge of The 2nd
Birmingham Battalion
£35

A rare other rank's
glengarry badge of The
Black Watch 3rd
Volunteer Batallion
£36

An officer's Waterloo
period universal pattern
shako plate. Of gilt die
struck copper   £36

A Commander in
Chief's Yeomanry
Escort slouch hat
badge   £40

A rare, other rank's
glengarry badge of The
Argyll and Sutherland
Highlanders   £40

Glengarry badge of
The 54th Regiment,
pre 1881.   £50

Cap badge of The
Alexandra P.W.O.
Hussars   £60

A cast brass Georgian crowned
oval pouch badge of The 15th
Regt. of Foot, circa 1775. £75

Officer's badge of
The North Devon-
shire Regiment
1822-29.   £90

good Victorian breastplate of heavy
orm, finely etched overall in foliage £36

A good, early 19th Century Household
Cavalry breastplate of heavy form £50

A good, mid-18th Century Cavalry
trooper's breastplate of heavy form
£52

A good, 17th Century breastplate of
Siege weight, pierced at top and edges
with four belt fastenings £62

An Imperial Russian Cavalry black
japanned cuirass and backplate. Both
edged with red cloth piping. £130

A good pair of late 16th Century
Italian breast and back plates, the
breast plate with pronounced medial
ridge and heavily roped neck. £205

A late 19th Century iron model of a traversing coastal defence gun, stepped, turned bottle barrel 17¾ ins. With brass trunnions, covers, capstan screw elevating , cascabel, rivetted sides, greased bed and rails for longitudinal movement and brass wheels on gantry for lateral movement £70

A Victorian model of an 18th Century field cannon. Brass barrel 14½ins. long. With iron shod wheels, Dolphin loops and wooden carriage. Overall length with carriage 23½ins. £102

A fine, late Georgian bronze barrelled signal cannon, barrel 22ins. With integrally turned re-inforcing rings, trunnions and cascabal and true dolphin carrying loops, the vent moulded as a shell. On its old stepped wooden carriage with wooden wheels.   £360

A large Victorian scale model of a 36 pounder heavy ordnance coastal defence gun, circa 1830. Heavy bronze barrel 23 ins. on stepped wooden carriage with brass wheels. Length of heavy slate base 38 ins. width 14½ ins.   £520

CLUBS

A rare, South Sea 'kite' shaped club, 27 ins. The haft part rattan bound
£26

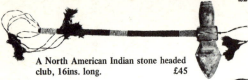

A North American Indian stone headed
club, 16ins. long.   £45

A North American Indian Plains club
I-Wata-Jinga, 21ins. long.   £75

CROSSBOWS

A late 18th Century stonebow 'prod' by Johnson of Wigan,
30 ins. overall span 25¼ ins. With an integral cocking lever
retained by sliding square, a husk engraved en suite with lock
£130

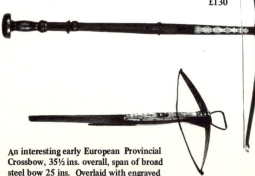

An interesting early European Provincial
Crossbow, 35½ ins. overall, span of broad
steel bow 25 ins. Overlaid with engraved
staghorn plates.   £250

Top: A rare, Georgian copper kettle drum of The Hon. East India Co.
On three legs. Diameter 19ins., height 19ins. £50

Bottom: A large, old Georgian military drum, with a wooden frame
and an old black painted finish. With cord stretchers. Diameter
20½ins., height 28½ins. £110

ELEPHANT GOADS

A good Indian elephant goad, Ankus, 20 ins. overall. The head with
traces of silver overlay, the wooden haft depicting animals £26

A very good Indian Elephant goad, Ankus,
the haft of ivory £75

FLASKS

A good, embossed copper gun flask, 'Shell'. £15

A copper powder flask, fluted. £15

A good, copper powder flask, 'Panel'. £20

A small copper priming flask, 3½ins. £26

A good, 17th century flattened horn powder flask, 4½ins. £28

A good copper powder flask 'Foliage'. £30

A leather covered powder flask from a cased rifle, 8½ins. With stamped patent top. £36

An unusual enamelled South Eurasian brass priming flask of circular form. £40

A copper bodied powder flask. Patent top stamped '1849 Batty'. £46

An embossed brass-mounted copper-bodied 3-way pistol flask. £51

An embossed copper powder flask, 8ins. £56

A nielloed silver mounted Caucasian priming flask, 5½ins. £58

An early 18th century flattened horn powder flask. £60

A fine Persian shell powder flask. £70

A German 17th century powder flask. Diameter 4¼ins. £75

A late 17th century flattened horn powder flask for a wheel-lock. £90

A 17th century, German musketeer's powder flask, 9½ins. £100

A 17th century musket priming flask of tri-angular shape, 3½ins tall. £155

GORGETS

Georgian officer's
copper gilt gorget.
£40

A Georgian officer's
gorget of The Cold-
stream Guards. Of
gilt copper. £43

A Georgian officer's
copper gilt universal
pattern gorget. £50

Late Georgian officer's
gorget of the 40th
Regiment. £65

A Georgian officer's
gorget, engraved with
crowned G.R. cypher.
£75

A George III officer's
gilt brass gorget. £85

Georgian officer's
copper gilt gorget
complete with
original chamois
lining. £90

A Georgian officer's
copper gilt gorget of
The Royal Marines.
£160

Georgian officer's
copper gilt gorget of
an Irish Volunteer
or Militia. £190

72

A Guatamalan military officer's kepi. Part leather covered with felt.   £15

An intricately woven cane helmet of the Nagas from Assam.   £20

A Prussian Infantryman's Picklehaube. With grey metal helmetplate and mounts.   £36

An officer's post-Crimea shako of the 47th Regt. of Foot.   £40

An other rank's busby of the 14th Hussars. With white hair plume.   £50

An officer's forage cap of the 1st V.B. Royal Fusiliers.   £54

A good, Indo-Persian steel helmet, Kulah Khud.   £60

An officer's bearskin of the Lancashire Fusiliers' Volunteer Battalion.   £61

A fireman's brass helmet with leather lined chin chains   £65

73

HELMETS

A French 19th century Cavalry helmet.   £65

A rare Nazi SS other ranks peaked cap. £66

A Victorian officer's cap of the Seaforth Highlanders.   £68

A well formed cabasset, forged from one piece of iron, circa 1600. £70

A 19th century Swedish cavalry leather helmet.   £80

Prussian Infantry Reservist officer's Picklehaube, gilt Helmet Plate.   £85

A Danish Dragoon, black leather helmet.   £90

Prussian pioneer Picklehaube of the 10th Pioneer Battalion, Helmet.   £90

A 17th century Burgonet of rather crude construction. £95

A Cromwellian lobster
tailed helmet.   £100

A good Indo-Persian
helmet Kulah Khud.
£100

An officer's 1869
pattern shako of
the Royal North
Gloucester Militia.
£110

An Indo-Persian
steel helmet
Kulah Khud. £110

Prussian artillery NCO's
Pickelhaube of the 46th
Field Artillery Regiment.
£135

An officer's Albert
shako of the 21st
(or 2nd West York)
Militia.   £135

A black leather helmet
of The King's Own
Norfolk Imperial
Yeomanry.   £160

A late 16th century
Burgonet, forged from
one piece of iron. £165

A scarce Prussian
cuirassier officer's
helmet with brass
chinscales.   £170

HELMETS

An Imperial Austrian Dragoon officer's helmet. £170

Victorian officer's shako of the Queen's Own Regiment of Yeomanry 1857-1861. £180

A good, late 16th century Italian Cabasset. £210

Rare other ranks 1842 helmet of the 6th Inniskilling Dragoons. £300

Imperial German Garde du Corps helmet. £310

Officer's black metal crested helmet of the Portland Legion, 1818-1834. £320

A rare early Victorian officers' lance cap of the 9th Lancers (Queens Royal). £550

An English close helmet, circa 1560-70. £600

A rare early 17th century close helmet, the skull formed in one piece. £720

A Victorian officer's post-1881 helmet plate of the Hampshire Regiment. £20

A helmet plate from The Royal West Kent Regiment.   £22

A Victorian officer's blue cloth helmet plate of The Essex Regiment.   £26

A Victorian officer's helmet of The Royal Wiltshire Militia. £29

A Victorian officer's silvered helmet plate of The Manchester Regt. 3rd Volunteer Battalion.   £30

A pre-1881 officer's blue cloth helmet plate of The 59th Regiment.   £30

Victorian officer's plate of The Devonshire Regiment, 1878.   £38

An officer's Albert Shako plate of The 51st (2nd Yorkshire West Riding) Light Infantry Regt. £40

A Victorian pre-1861 officer's blue cloth helmet plate of The 46th Regiment.   £52

A helmet plate of The Lothian Regiment. £55

An officer's shako plate 1812-16-pattern of The 21st (Royal North British Fusiliers) Regiment. £85

Officer's plate of The 16th Bedfordshire Regiment.   £175

77

MISCELLANEOUS
ALARM GUNS

A scarce, 4-shot Naylor's Patent percussion repeating alarm gun, 6¾ins. by 2¾ins. by 1½ins.   Numbered 2323.   £32

A most unusual, tube or teat ignited alarm gun, cast iron base 11ins. With a hinged trigger with holes for trip wire and a brass friction roller bearing hammer which fall across the tube or teat of the alarm cartridge.   £36

KEY PISTOLS

A very well made percussion key pistol, circa 1850 in style, 7½ins. long.   £140

An extremely well made flintlock key pistol in the French style of circa 1640, 11ins. long.   £270

JINGASA

A good Japanese Jingasa. The surface decorated with mother-of-pearl inlaid decoration and a Satsuma mon.   £40

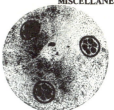

A good Japanese Jingasa. Covered overall with gilt lacquer decoration. £70

TANJU AIKUCHI

A good scarce 28-bore Japanese Tanju Aikuchi, 15½ins. long. £200

TELESCOPES

A naval officer's telescope, circa 1750, 42ins. overall. Ten sided wooden frame takes down into four sections for packing in a sea chest. With sliding lens covers.   £58

A German silver telescope, presented by King Wilhelm in 1875, 32¾ins. extended. With extending lens hood with white metal dust cover. Covered in black leather and contained in its rosewood veneered presentation case which bears presentation inscription.   £65

SABRETACHES

A rare Victorian officer's brown leather sabretache of The Portsmouth Civil Service Corps.   £36

A Victorian officer's full-dress sabretache of The Royal Horse Artillery   £90

A Victorian Royal Artillery officer's sabretache   £80

An officer's full-dress sabretache of The 15th Hussars   £150

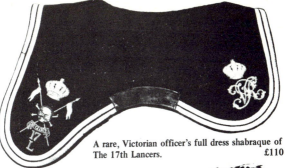

A rare, Victorian officer's full dress shabraque of
The 17th Lancers. £110

A Victorian officer's leopard skin shabraque of
The 15th Hussars. £135

A fine, post-1902 officer's shabraque of
The First Life Guards. £180

SHOULDER BELT PLATES

A post-1902 officer's rectangular shoulder belt plate of The Queen's Own Cameron Highlanders £14

Other rank's brass plate of The Wiltshire Regiment, circa 1850. £28

Victorian officer's plate of The King's Own Light Infantry Militia £34

A Georgian other rank's cast brass oval shoulder belt plate. Bearing initials 'WSM' £35

Officer's plate of The Durham Light Infantry 1830-55 £40

An officer's shoulder belt plate of The Royal Welsh Fusiliers circa 1840-55 £42

Georgian officer's plate with the motto 'Regi Patriae Que Fidelis'. £44

Belt plate of The Grenadier Guards. Pre 1855 issue £46

Officer's plate of The South Devon 1844-45 £50

Georgian shoulder belt plate of The Letterkenny Corps. £55

A fine Victorian rectangular gilt shoulder belt plate of The Royal Marines £90

Early 19th Century officer's plate of The Scots Guards £115

A George VI full dress bullion embroidered pouch and gold lace belt of The Corps of Gentlemen at Arms. The belt with plain burnished brass fittings £30

A post-1902, officer's full dress bullion embroidered pouch and belt of The 7th Queen's Own Hussars. The belt with gilt fittings £48

A Victorian officer's silver fronted pouch of The 2nd West York Yeomanry. The hallmarked silver front with engraved border and superimposed silver and gilt device. On a hallmarked gold lace belt with hallmarked silver buckle, prickers, chains etc. £60

An officer's silver fronted pouch of The Hyderabad Contingent Lancers. With a silver front hallmarked Birmingham 1905 with engraved border and superimposed gilt device with late Victorian crown. With its gold lace belt and matching silver mounts (1903) including crown, chains, prickers etc. £60

SPURS

A good quality nickel silver inlaid South American iron spur £11

One of a pair of South American rowel spurs. With nickel overlaid bows and massive pierced rowel bridles £40

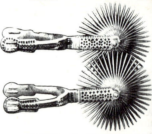

A good pair of South American rowel spurs. With large nickel silver buckles, iron bows chiselled on one side and inlaid with scroll engraved nickel silver on the other £42

An ornate pair of South American German silver gaucho spurs £60

SPORRANS

A fine Scottish hallmarked silver mounted sporran. £36

A good officer's sporran of The Seaforth Highlanders £38

An officer's full dress sporran of The Black Watch £40

A rare, 18th Century tinder box
4½ ins. overall   £38

A flintlock 'cottage' tinderlighter,
7 ins. With a brass frame and box
with door stamped 'Hinde'   £50

A mid-18th Century flintlock 'cottage'
tinderlighter, 7ins. With a brass frame
and box with door stamped 'Hinde'. £95

A scarce, all brass British military
flintlock boxlock cannon igniter,
circa 1800, 5¼ ins.   £115

TRUNCHEONS

A good William IV painted truncheon, 18 ins. long   £20

An interesting Victorian painted truncheon bearing the
crowned arms of The London and Birmingham Railway   £40

An interesting old truncheon made from a whale bone marlin
spike, 17 ins. long   £60

TSUBAS

A Japanese oval silver dagger tsuba, engraved with a simple pattern in katakiri £21

A good Japanese iron tsuba showing a chidari perched on a twig, possibly 18th Century goto work £21

A good Japanese iron tsuba, possibly 18th century mito-work £31

A large iron tsuba, chiselled in low relief with a man bearing a sword Ken who has frightened an oni £44

A good Japanese open-work iron tsuba, pierced with stylised bamboo and foliage, signed 'Coshu No Ju Massasaba' £50

Mid 18th century tsuba chiselled in the round with a dragon amidst foliage and waves £55

Iron tsuba decorated with tangs £160

Early Edo period iron tsuba depicting Shoki £500

19th century Ishigura school tsuba depicting a cockerel strutting through some flowers £2,600

A superb Victorian footman's livery of
The Earls of Ashburnham £40

A Georgian costume of blue velvet worn
ceremoniously at Eton College £30

A late Georgian military long-tailed
scarlet coatee of The 18th Regiment
of Foot £45

Victorian officer's scarlet coatee of
The lst Dorset Militia, circa 1845 £75

A fine page of honour's uniform as worn at the 1911 Coronation. £100

A fine and rare officer's full dress scarlet jacket of the 1st Bombay Native Cavalry, 1819. £125

A complete State Trumpeter's uniform of the 6th (Inniskilling) Dragoons and the 14th/20th Hussars. £250

A complete Victorian officer's uniform of the 12th Prince of Wales's Lancers. Including six battle honours. £350

An Indo-Persian bi-dent head, wavy blades 12ins., with reinforced points. With a metal half mount with traces of silver damascus decoration.   £27

An Indo-Persian Trident head.   The surface covered in inscriptions within brass borders, the inscriptions overlaid in silver and brass.   With a metal haft also covered in inscriptions and similarly decorated.   £30

A good, early 19th century engraved copy of a 17th century Saxon Parade Halberd, overall length 7ft 3ins., overall height of head 21½ins. with base. With a broad spike with central rib and etched side straps.   Dated 'MDCXI'.
                                                                                                £52

A good, 16th century Italian polearm coresque or runkas, head 31ins., auxilliary blades 7ins. With an 18th century velvet covered haft and neck with bullion embroidered tassels.   £52

A Persian Trident head, overall length 23ins.   Central blade bordered with two flamberge blades, all with reinforced points.   With a metal haft mount.   £54

89

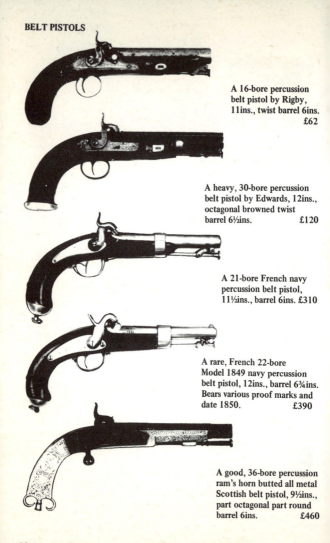

A 16-bore percussion belt pistol by Rigby, 11ins., twist barrel 6ins. £62

A heavy, 30-bore percussion belt pistol by Edwards, 12ins., octagonal browned twist barrel 6½ins. £120

A 21-bore French navy percussion belt pistol, 11½ins., barrel 6ins. £310

A rare, French 22-bore Model 1849 navy percussion belt pistol, 12ins., barrel 6¾ins. Bears various proof marks and date 1850. £390

A good, 36-bore percussion ram's horn butted all metal Scottish belt pistol, 9½ins., part octagonal part round barrel 6ins. £460

A good, 16-bore percussion Yeomanry carbine, 39ins., twist barrel 20ins. £100

A 29-bore breech loading percussion carbine, Cooper's Patent action, 48½ins., barrel 29ins. to breech. £110

A fine, 13-bore French An 22 military percussion cavalry carbine, 34½ins., barrel 20ins. Halfstocked, with regulation lock, brass mounts and dated 1847. £135

A good, 44-bore breech loading Westley Richard's Patent monkey tail military percussion carbine, 36ins., barrel 20ins. £170

A fine, 44-bore desirable sealed pattern breech loading Westley Richards Patent monkey tail military percussion carbine. 36ins., barrel 20ins. £420

A good, 18-bore East India Company military percussion
cavalry holster pistol, 14ins., barrel 8ins.   £170

A good, 18-bore East India Company military percussion
cavalry holster pistol, 14ins., barrel 8ins.   £200

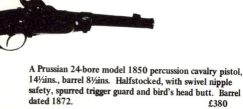

A Prussian 24-bore model 1850 percussion cavalry pistol,
14½ins., barrel 8½ins. Halfstocked, with swivel nipple
safety, spurred trigger guard and bird's head butt. Barrel
dated 1872.   £380

A very fine and desirable 25-bore Enfield rifled percussion
cavalry pistol, 15½ins., barrel 10ins. Tower proved, with
small leaf sight to 100yds., deep grooved rifling, steel
lanyard ring and swivel mounts.   £510

A scarce, 6-shot, 7.65mm. Belgian double action dagger revolver, 7 ins., octagonal barrel 3½ins.   £70

A rare, 6-shot pinfire Apache combination knuckleduster revolver, 9ins. long extended.   £220

A good and rare double barrelled 80-bore, percussion dagger pistol.   £220

A very fine old flintlock hunting dirk pistol, single edged blade, 14ins. long.   £240

A scarce, .41in. rimfire Remington over and under Derringer, 4¾in., barrels 3in.   £40

A scarce, .41 in. Remington No. 72. With ivory grips.   Plated overall.   £46

A scarce, .41 in. rimfire Derringer No. 656. £66

A rare, five barrelled .22in. Remington ring trigger Derringer, 4¾in., fluted barrels 3in.   £82

A rare .41 Moore's pattern 1863 presentation Derringer, side swing plated barrel 3in.   £140

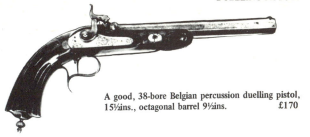

A good, 38-bore Belgian percussion duelling pistol, 15½ins., octagonal barrel 9½ins.   £170

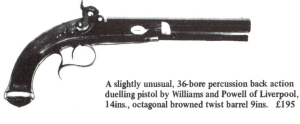

A slightly unusual, 36-bore percussion back action duelling pistol by Williams and Powell of Liverpool, 14ins., octagonal browned twist barrel 9ins.   £195

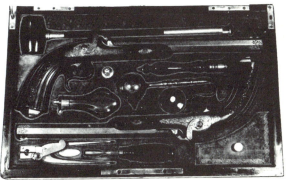

A pair of 36-bore Cont. percussion duelling pistols, 17ins., octagonal rifled barrels 11ins. Scroll engraved at breeches and muzzles, half-stocked in scroll carved and fluted ebony with scroll engraved steel mounts and detented locks.   £1,750

EPROUVETTES (POWDER TESTERS)

An interesting, hand ignited eprouvette by G. and J Hawksley, 7¼ins. overall, barrel 1½ ins. Of lacquered brass with steel wheel and turned ebonised handles. £80

An unusually fine and large brass framed flintlock boxlock eprouvette 6½in. overall by Monks

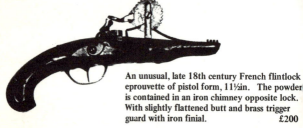

An unusual, late 18th century French flintlock eprouvette of pistol form, 11½in. The powder is contained in an iron chimney opposite lock. With slightly flattened butt and brass trigger guard with iron finial. £200

An unusual, all steel Cont. flintlock eprouvette, 9¼in. Exposed flintlock action. The powder is contained in a tube below the pan. £260

A scarce, .25in rimfire Unwin and Rodgers patent knife pistol, 6½ ins., white metal octagonal barrel 3¾ins. With a white metal frame with cartridge trap door, folding trigger with a 3½ins bowie shaped blade and 3¾ins straight blade. Birmingham Black Powder proved.   £64

A scarce, .30 Unwin and Rodgers percussion knife pistol, 6½ ins., German silver barrel 3¼ ins. Birmingham proved. Containing tweezers and mould, two blades and folding trigger. With German silver mounts and frame horn sideplates.   £150

A scarce, double barrelled 80-bore side by side boxlock Cont. percussion knife pistol, 14½ins., single fullered double edged blade 10ins., barrels 3¼ ins. With folding trigger and hammers formed as part of the crosspiece.   £180

A scarce, 7.65mm. centre fire Cont. knife pistol combination Presentation knife 5in. (closed). Blade 3¼in. With hinge up breech block, cork-screw acts as a trigger.   £240

HOLSTER PISTOLS

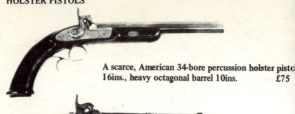

A scarce, American 34-bore percussion holster pistol 16ins., heavy octagonal barrel 10ins. £75

An American 28-bore military percussion holster pistol 14ins., barrel 8½ins. £85

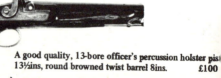

A large, 15-bore , East India Company military percussion holster pistol, 15½ins., barrel 9ins. £90

A good quality, 13-bore officer's percussion holster pistol 13½ins, round browned twist barrel 8ins. £100

A French, 13-bore Model 1777 military percussion Cavalry holster pistol, 13ins., barrel 7ins. £135

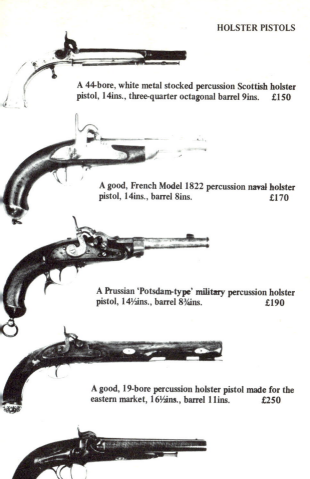

A 44-bore, white metal stocked percussion Scottish holster pistol, 14ins., three-quarter octagonal barrel 9ins.   £150

A good, French Model 1822 percussion naval holster pistol, 14ins., barrel 8ins.   £170

A Prussian 'Potsdam-type' military percussion holster pistol, 14½ins., barrel 8¾ins.   £190

A good, 19-bore percussion holster pistol made for the eastern market, 16½ins., barrel 11ins.   £250

A very fine, French double barrelled 19-bore side hammer percussion holster pistol, 14ins., twist steel barrels 7ins.   £310

99

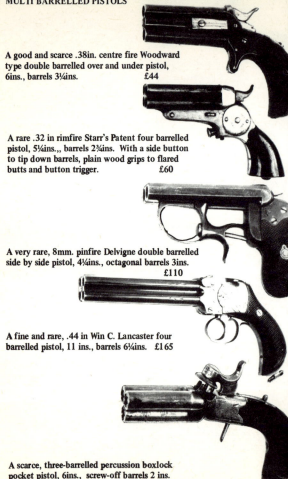

A good and scarce .38in. centre fire Woodward
type double barrelled over and under pistol,
6ins., barrels 3¼ins.   £44

A rare .32 in rimfire Starr's Patent four barrelled
pistol, 5¼ins.,, barrels 2¾ins. With a side button
to tip down barrels, plain wood grips to flared
butts and button trigger.   £60

A very rare, 8mm. pinfire Delvigne double barrelled
side by side pistol, 4¼ins., octagonal barrels 3ins.
                                               £110

A fine and rare, .44 in Win C. Lancaster four
barrelled pistol, 11 ins., barrels 6¼ins.   £165

A scarce, three-barrelled percussion boxlock
pocket pistol, 6ins., screw-off barrels 2 ins.
Birmingham proved.   £202

A rare .50/70 Springfield musket, 47½ins., barrel 28¾ins. Converted from percussion to centre fire during U.S. Civil War. £60

A 14-bore French military percussion musket, 45ins., barrel 29¾ins. With Birmingham proved, fullstocked lockplate and sling swivels. £135

A good Japanese matchlock musket, 52ins., octagonal barrel 39½ins. With fullstocked brass mounts, brass lockplate, swivel pan cover and breech band engraved with kabuto and mask. Overlaid with silver. £140

A 32-bore Japanese matchlock musket, 51½ins., octagonal barrel 38½ins. Fullstocked in cherry wood with brass pan cap and lock and well formed butt plate. £140

PEPPERBOX REVOLVERS

A rare 6-shot, .22in Bacon Arms Co
pepperbox revolver, 5½ins., barrels
2½ins.   £50

A scarce, 6-shot, .25 ins pinfire pepperbox
double action revolver, 4¾ins., barrels 1¾in
£95

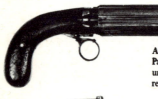

An unusual, 6-shot, 60-bore Cooper
Patent double action ring-trigger
underhammer percussion pepperbox
revolver, 7½ins., barrels 3ins.   £115

A good, 6-shot, 90-bore top snap
percussion pepperbox revolver, 8in
barrels 3ins.   £130

A 6-shot, 54-bore top snap self cockin
percussion pepperbox revolver, 9½ins.
fluted barrels 4ins. Birmingham prove
£1

A good, French four barrelled understriker pepperbox revolver, 7½ins., barrels 3½ins. With scroll engraved rounded frames and ring trigger. £165

A rare and desirable, 4-shot 11mm. pinfire Le Faucheux underhammer pepperbox revolver, 8ins., barrels 3ins. £190

A rare, 5-shot 28 Robbin and Lawrence double action percussion pepperbox revolver, 7½ ins., barrel block 3½ ins. The front 2¼ ins. of the barrels unscrew and slide forward to facilitate loading the chambers. £230

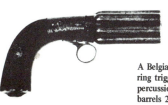

A Belgian, six-barrelled, 74-bore Mariette ring trigger underhammer double action percussion pepperbox revolver, 7¼ins., barrels 2¾ins. £240

A very rare, 5-shot, .30 ins. Budding percussion pepperbox revolver, 8ins., bronze barrels 3¼ ins. £700

103

A rare, .38in. Colt Brownings Patent 1900 auto pistol, 9ins., barrel 6ins. With spur hammer and rear sight safety. Dated 1897.   £21

A scarce, 7.65mm. Steyr-Mannlicher 1905 semi auto pistol, 9½ins., barrel 6¼ins. With lanyard loop, lined wood grips and external hammer.   £51

A rare, 7.63mm. Luger 1900-1906 commercial auto pistol, 9ins., barrel 4¾ins. With grip safety and plain side safety.   £70

A rare, 7.63mm. Luger 1900-1906 commercial auto pistol, 9ins., barrel 4¾ins. With plain side safety and dished toggle thumb pieces.   £165

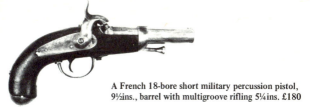

A French 18-bore short military percussion pistol,
9½ins., barrel with multigroove rifling 5¼ins. £180

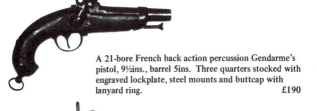

A 21-bore French back action percussion Gendarme's
pistol, 9½ins., barrel 5ins. Three quarters stocked with
engraved lockplate, steel mounts and buttcap with
lanyard ring. £190

A fine, 13-bore French An 22 military percussion
pistol, 14ins., barrel 8ins. Halfstocked, with
engraved lockplate. £230

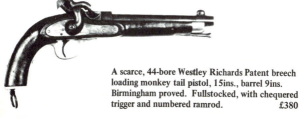

A scarce, 44-bore Westley Richards Patent breech
loading monkey tail pistol, 15ins., barrel 9ins.
Birmingham proved. Fullstocked, with chequered
trigger and numbered ramrod. £380

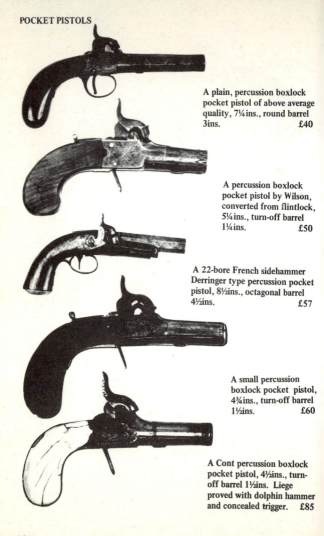

A plain, percussion boxlock pocket pistol of above average quality, 7¼ ins., round barrel 3ins. £40

A percussion boxlock pocket pistol by Wilson, converted from flintlock, 5¼ ins., turn-off barrel 1¼ ins. £50

A 22-bore French sidehammer Derringer type percussion pocket pistol, 8½ins., octagonal barrel 4½ins. £57

A small percussion boxlock pocket pistol, 4¾ins., turn-off barrel 1½ins. £60

A Cont percussion boxlock pocket pistol, 4½ins., turn-off barrel 1½ins. Liege proved with dolphin hammer and concealed trigger. £85

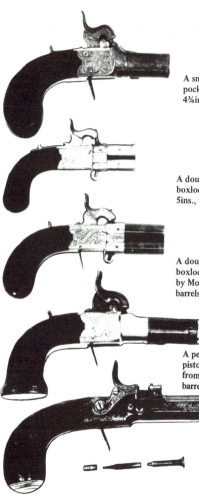

A small percussion boxlock pocket pistol by Collins, 4¾ins., turn-off barrel 1¼ins. £100

A double barrelled percussion boxlock turnover pocket pistol, 5ins., turn-off barrels 1in. £105

A double barrelled percussion boxlock turnover pocket pistol by Monks, 6ins., turn-off barrels 1½ins. £135

A percussion boxlock pocket pistol by Goodwin, converted from flintlock, 6ins., turn-off barrel 1½ins. £140

A rare, 38-bore Forsyth's Patent fulminate percussion system pocket pistol, No. 3222, circa 1825, 6½ins., octagonal barrel 2¾ins. £720

A well made six shot 12mm French
Le Faucheux type double action
pinfire revolver 10¼ins., octagonal
barrel 5¾ins.   £40

A good, 5-shot, .38ins. rimfire Tranters
Patent double action revolver, 10½ins.,
octagonal barrel 5ins.   £54

A most interesting, 5-shot 11mm Vero
Montenegro military style double action
revolver, 10ins., barrel 5ins.   £64

A good, scarce, 6-shot, 8mm rimfire
'Mariette Brevete' open top single
action revolver, 11½ins., octagonal
barrel 6½ins.   £65

An unusual, 12-shot Cont.9mm pinfire
double action revolver, 11ins., barrel 6ins.
With an open frame, chequered wood grips
and a lanyard ring to butt.   £70

108

A good and scarce, 6-shot, .45in. centre fire 'Hills Patent' self extracting double action revolver, 10½in., octagonal barrel 6in. £75

A scarce, 5-shot, .50ins. Tranters Patent double action revolver, 9¾ins., octagonal barrel 4½ins.
£80

A Cont, copy of a 5-shot, 54-bore Beaumont Adams double action percussion revolver, 11ins., barrel 6ins. £85

An interesting, 5-shot, .45 self cocking Adams Patent model 1851 percussion revolver, 12½ins., octagonal barrel 6¾ins. £100

A rare, 5-shot, .32ins. centre fire Belgian long cylinder double action revolver, 8¾ins., octagonal barrel 3ins. £140

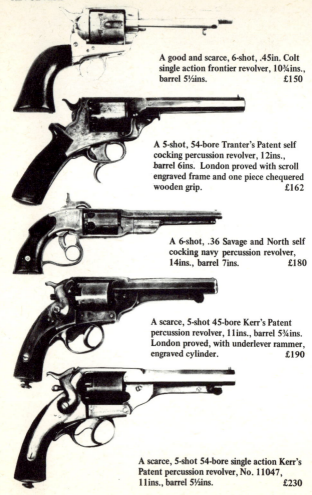

A good and scarce, 6-shot, .45in. Colt single action frontier revolver, 10¾ins., barrel 5½ins. £150

A 5-shot, 54-bore Tranter's Patent self cocking percussion revolver, 12ins., barrel 6ins. London proved with scroll engraved frame and one piece chequered wooden grip. £162

A 6-shot, .36 Savage and North self cocking navy percussion revolver, 14ins., barrel 7ins. £180

A scarce, 5-shot 45-bore Kerr's Patent percussion revolver, 11ins., barrel 5¾ins. London proved, with underlever rammer, engraved cylinder. £190

A scarce, 5-shot 54-bore single action Kerr's Patent percussion revolver, No. 11047, 11ins., barrel 5½ins. £230

A 5-shot, 54-bore 1851 model Adams self cocking percussion revolver, 12½ins., barrel 6¼ins. With chequered butt and engravings on top strap and frame. In its velvet lined oak case.   £340

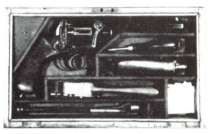

A good, 5-shot 54-bore self cocking Adams Patent percussion revolver, 11½ins., barrel 6¼ins. London proved, with spring safety catch and chequered one piece wood grips. Contained in green baize lined oak fitted case.   £485

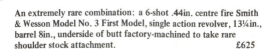

An extremely rare combination: a 6-shot .44in. centre fire Smith & Wesson Model No. 3 First Model, single action revolver, 13¼in., barrel 8in., underside of butt factory-machined to take rare shoulder stock attachment.   £625

A 6-shot, .44in. Colt Army four-screw
Model 1860 single action percussion
revolver, 14ins., round barrel 8ins. £33

A 5-shot, .36in. Manhattan single action
percussion revolver, 9½ins., barrel 4ins. £55

A 5-shot, .36in. Manhattan single action
percussion pocket revolver, 9ins., barrel
4ins. With engraved cylinder and brass
backstrap and trigger. £80

A scarce, 6-shot, .36 Beals Model Remington
single action percussion revolver No. 16893,
13½ins., octagonal barrel 7½ins. to cylinder.
£110

A 6-shot, .36in. single action Colt navy percussion
revolver No. 108742, 13½ins., octagonal barrel
7½ins. £130

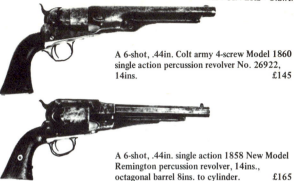

A 6-shot, .44in. Colt army 4-screw Model 1860 single action percussion revolver No. 26922, 14ins.   £145

A 6-shot, .44in. single action 1858 New Model Remington percussion revolver, 14ins., octagonal barrel 8ins. to cylinder.   £165

A 6-shot, .36in. single action Colt navy percussion revolver, 13ins., barrel 7½ins.   £200

An interesting, 6-shot, .44in. Colt army single action percussion revolver, 13½ins., barrel 8ins.   £250

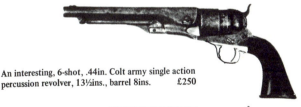

A 6-shot, .31in. Presentation Colt single action revolver, 10ins., octagonal barrel 5ins. With Presentation inscription and engraved cylinder.   £310

113

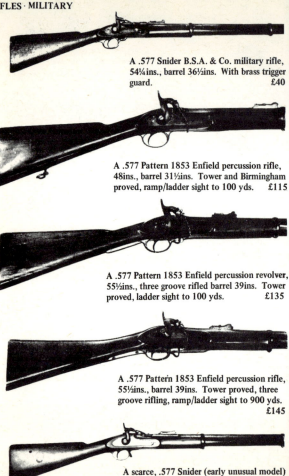

A .577 Snider B.S.A. & Co. military rifle, 54¼ins., barrel 36½ins. With brass trigger guard. £40

A .577 Pattern 1853 Enfield percussion rifle, 48ins., barrel 31½ins. Tower and Birmingham proved, ramp/ladder sight to 100 yds. £115

A .577 Pattern 1853 Enfield percussion revolver, 55½ins., three groove rifled barrel 39ins. Tower proved, ladder sight to 100 yds. £135

A .577 Pattern 1853 Enfield percussion rifle, 55½ins., barrel 39ins. Tower proved, three groove rifling, ramp/ladder sight to 900 yds. £145

A scarce, .577 Snider (early unusual model) 1864 military rifle, 55¼ins., barrel 36½ins. £145

A rare, .577 Storm's Patent breech loading
percussion rifle, 48½ins., barrel 30½ins.
Fullstocked, with a sideways-swivelling
breech.   £150

A .577 three band Volunteer Enfield
percussion rifle, 55½ins., barrel 39ins.
Ladder rear sight to 900 yds.   £150

A French 12mm breech loading military
percussion rifle, 46½ins., barrel 30ins.
With bayonet lug below muzzle.   £190

A double barrelled .577 Jacobs military
percussion rifle, 40½ins., barrels 24ins.
With four groove Jacobs rifling.   £250

A good, scarce, .451 Kerr's Patent percussion
target rifle by the London Armoury Company,
53¼ins., barrel 37ins.   London proved, ladder
sight to 1,000 yds.   £320

A well made, .450in. Martini Henry Patent rifle
by Field of London, 46ins., barrel 29ins.   £48

A rare, .45/70 Springfield Model 1878 military
rifle, 51½ins., barrel 32½ins.   £70

A scarce, .45in. (1½in. case) Westley Richards military
rifle, 40¼ins., barrel 23¾ins.   Birmingham Black
Powder proved.   £80

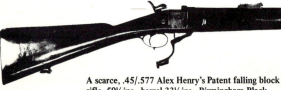

A scarce, .45/.577 Alex Henry's Patent falling block
rifle, 50¼ins., barrel 33¼ins.   Birmingham Black
Powder proved.   £85

A rare, .520in. centre fire Wilson's Patent Sea Service
rifle, 49ins., barrel 30½ins.   Birmingham Black Powder
proved.   £105

A rare, .577in. Allin (or Beran's) Patent trapdoor breech load rifle, 55¼ins., barrel 36¼ins. Birmingham Black Powder proved. £125

A fine, Indian Mik gun Torador, 67ins., octagonal barrel 50ins. Finely decorated at muzzle and breech and along edges with gold damascus foliate patterns. £180

A fine and scarce, .45/60in. Winchester Model 1876 sporting rifle, 47ins., half round half octagonal barrel 26ins. Adjustable leaf backsight to 1,000 yds. £185

A rare, 64-bore, 9-shot Porter's Patent repeating percussion turret rifle, 43ins., heavy octagonal barrel 22½ins. £280

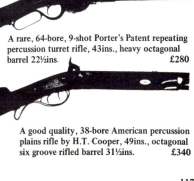

A good quality, 38-bore American percussion plains rifle by H.T. Cooper, 49ins., octagonal six groove rifled barrel 31½ins. £340

SALOON PISTOLS

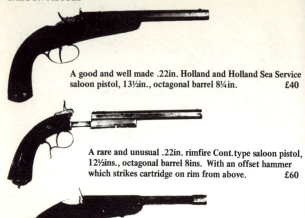

A good and well made .22in. Holland and Holland Sea Service
saloon pistol, 13½in., octagonal barrel 8¼in.   £40

A rare and unusual .22in. rimfire Cont.type saloon pistol,
12½ins., octagonal barrel 8ins. With an offset hammer
which strikes cartridge on rim from above.   £60

A fine and well made .22in. (B.B. cap) Cont. saloon pistol
14in., octagonal barrel 8in.   With a one piece stock, capped
butt and spur trigger guard.   £72

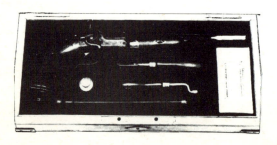

An extremely fine and well made .44in. Gastinne Renette saloon pistol,
16½ins., half octagonal and half round barrel 10ins. With a plain wood
forend secured by wedge.   £520

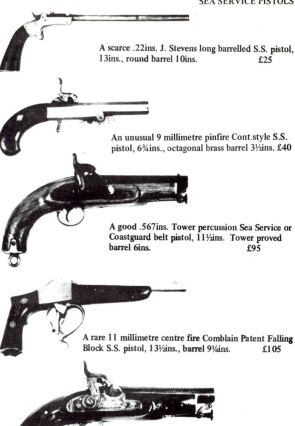

A scarce .22ins. J. Stevens long barrelled S.S. pistol, 13ins., round barrel 10ins. £25

An unusual 9 millimetre pinfire Cont.style S.S. pistol, 6¾ins., octagonal brass barrel 3½ins. £40

A good .567ins. Tower percussion Sea Service or Coastguard belt pistol, 11½ins. Tower proved barrel 6ins. £95

A rare 11 millimetre centre fire Comblain Patent Falling Block S.S. pistol, 13½ins., barrel 9¼ins. £105

A 17-bore percussion Sea Service Coastguard-type pistol by Beckwith, 11½ins., barrel 6ins. With plain fullstock and butt and swivel ramrod. £115

119

A French, 13-bore Model 1822 percussion naval pistol, 14ins., barrel 8ins. £145

A rare .50in. Remington Rolling Block S.S. pistol, 13¼in., barrel 8½in., London Black Powder proved. £160

A French, 13-bore Model 1822 percussion naval pistol, 14ins., barrel 8ins. £175

A very rare .557 centre fire hinge breech Adam's Patent S.S. pistol, 16ins., barrel 8¾ins. £202

A rare, French 22-bore Model 1837 percussion naval belt pistol, 11½ins., barrel 6ins. £310

A most unusual, percussion sporting gun, 24½ins., wooden barrel 12½ins. Dated 1840, this was probably made by a gunsmith for a child. £56

A fine, 18-bore percussion sporting gun, by I. Smith, converted from flintlock by the maker, 50ins., blued barrel 34ins. £125

A good 18-bore double barrelled back action German percussion sporting gun, 48ins. £240

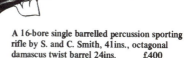

A 16-bore single barrelled percussion sporting rifle by S. and C. Smith, 41ins., octagonal damascus twist barrel 24ins. £400

A most desirable top quality Hallmarked silver mounted percussion sporting rifle by Samuel and C. Smith. £540

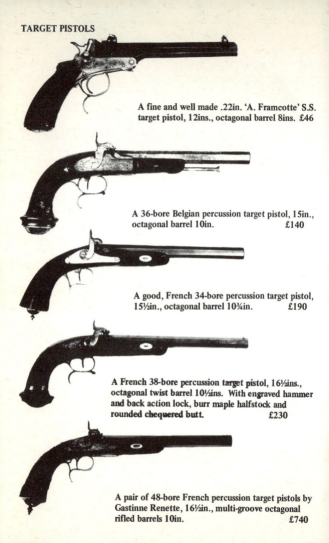

A fine and well made .22in. 'A. Framcotte' S.S.
target pistol, 12ins., octagonal barrel 8ins. £46

A 36-bore Belgian percussion target pistol, 15in.,
octagonal barrel 10in.   £140

A good, French 34-bore percussion target pistol,
15½in., octagonal barrel 10¾in.   £190

A French 38-bore percussion target pistol, 16½ins.,
octagonal twist barrel 10½ins. With engraved hammer
and back action lock, burr maple halfstock and
rounded chequered butt.   £230

A pair of 48-bore French percussion target pistols by
Gastinne Renette, 16½in., multi-groove octagonal
rifled barrels 10in.   £740

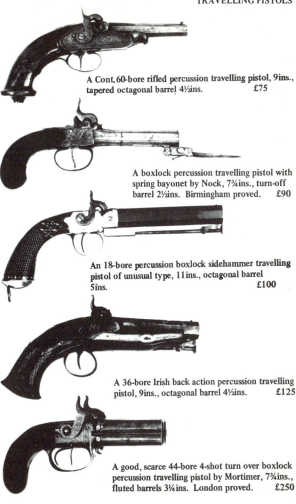

A Cont. 60-bore rifled percussion travelling pistol, 9ins., tapered octagonal barrel 4½ins.   £75

A boxlock percussion travelling pistol with spring bayonet by Nock, 7¾ins., turn-off barrel 2½ins. Birmingham proved.   £90

An 18-bore percussion boxlock sidehammer travelling pistol of unusual type, 11ins., octagonal barrel 5ins.   £100

A 36-bore Irish back action percussion travelling pistol, 9ins., octagonal barrel 4½ins.   £125

A good, scarce 44-bore 4-shot turn over boxlock percussion travelling pistol by Mortimer, 7¾ins., fluted barrels 3¼ins. London proved.   £250

INDEX

124